D0946939

Dinosaurs Alive!

Tyrannosaurus

and Other Mighty Hunters

Jinny Johnson

Illustrated by Graham Rosewarne

A+
Smart Apple Media

Published by Smart Apple Media
2140 Howard Drive West
North Mankato, MN 56003

Designed by Helen James
Edited by Mary-Jane Wilkins
Artwork by Graham Rosewarne

Printed in China

Library of Congress Cataloging-in-Publication Data

Johnson, Jinny.
Tyrannosaurus and other mighty hunters / by Jinny Johnson.
p. cm. — (Dinosaurs alive!)
Includes index.
ISBN 978-1-59920-063-7
1. Tyrannosaurus rex—Juvenile literature. 2. Predatory animals—Juvenile literature. I. Title.

QE862.S3J64 2007
567.912'9—dc22 2006102832

First Edition

9 8 7 6 5 4 3 2 1

Contents

A dinosaur's world

A dinosaur was a kind of reptile that lived millions of years ago. Dinosaurs lived long before there were people on Earth.

We know about dinosaurs because many of their bones and teeth have been discovered. Scientists called paleontologists (pay-lee-on-ta-loh-jists) learn a lot about the animals by studying these bones.

The first dinosaurs lived about 225 million years ago. They disappeared—became extinct—about 65 million years ago. Some scientists believe that birds are a type of dinosaur, so they say there are still dinosaurs living all around us!

Spinosaurus

TRIASSIC

248 to 205 million years ago

Some dinosaurs that lived at this time:
Coelophysis, Eoraptor, Liliensternus, Plateosaurus, Riojasaurus, Saltopus

EARLY JURASSIC

205 to 180 million years ago

Some dinosaurs that lived at this time:
Crylophosaurus, Dilophosaurus, Lesothosaurus, Massospondylus, Scelidosaurus, Scutellosaurus

LATE JURASSIC

180 to 144 million years ago

Some dinosaurs that lived at this time:
Allosaurus, Apatosaurus, Brachiosaurus, Ornitholestes, Stegosaurus, Diplodocus

Diplodocus

EARLY CRETACEOUS

144 to 98 million years ago

Some dinosaurs that lived at this time: Baryonyx, Giganotosaurus, Iguanodon, Leaellynasaura, Muttaburrasaurus, Nodosaurus, Sauropelta

LATE CRETACEOUS

98 to 65 million years ago

Some dinosaurs that lived at this time:
Ankylosaurus, Gallimimus, Maiasaura, Triceratops, Tyrannosaurus, Velociraptor

Velociraptor

Fiercest of all

The tyrannosaurus was one of the largest and fiercest of all dinosaurs—and one of the biggest meat-eating animals that has ever lived.

This mighty killer had a large, powerful head, a strong body, and a heavy tail. Its back legs were big and strong, but its arms were too short even to reach its mouth!

The tyrannosaurus hunted other dinosaurs —usually large, slow-moving plant-eaters. Its jaws and teeth were so strong it could even crunch up bones.

The tyrannosaurus walked upright on its back legs. It held its tail straight out behind to help balance the weight of its enormous head.

This is how you say tyrannosaurus: tie-ran-oh-sore-us

TYRANNOSAURUS

Group: tyrannosaurids

Length: up to 39 feet (12 m)

Lived in: North America

When: Late Cretaceous, 67–65 million years ago

Dinosaurs lived long before there were people on Earth. But here you can see how big a dinosaur was compared to a seven-year-old child.

Inside a tyrannosaurus

If the tyrannosaurus were alive today, it would tower over the world's largest hunting animals, such as tigers and bears.

An animal this size needed a big skeleton that was strong enough to support its huge body, but not so heavy that it couldn't move fast.

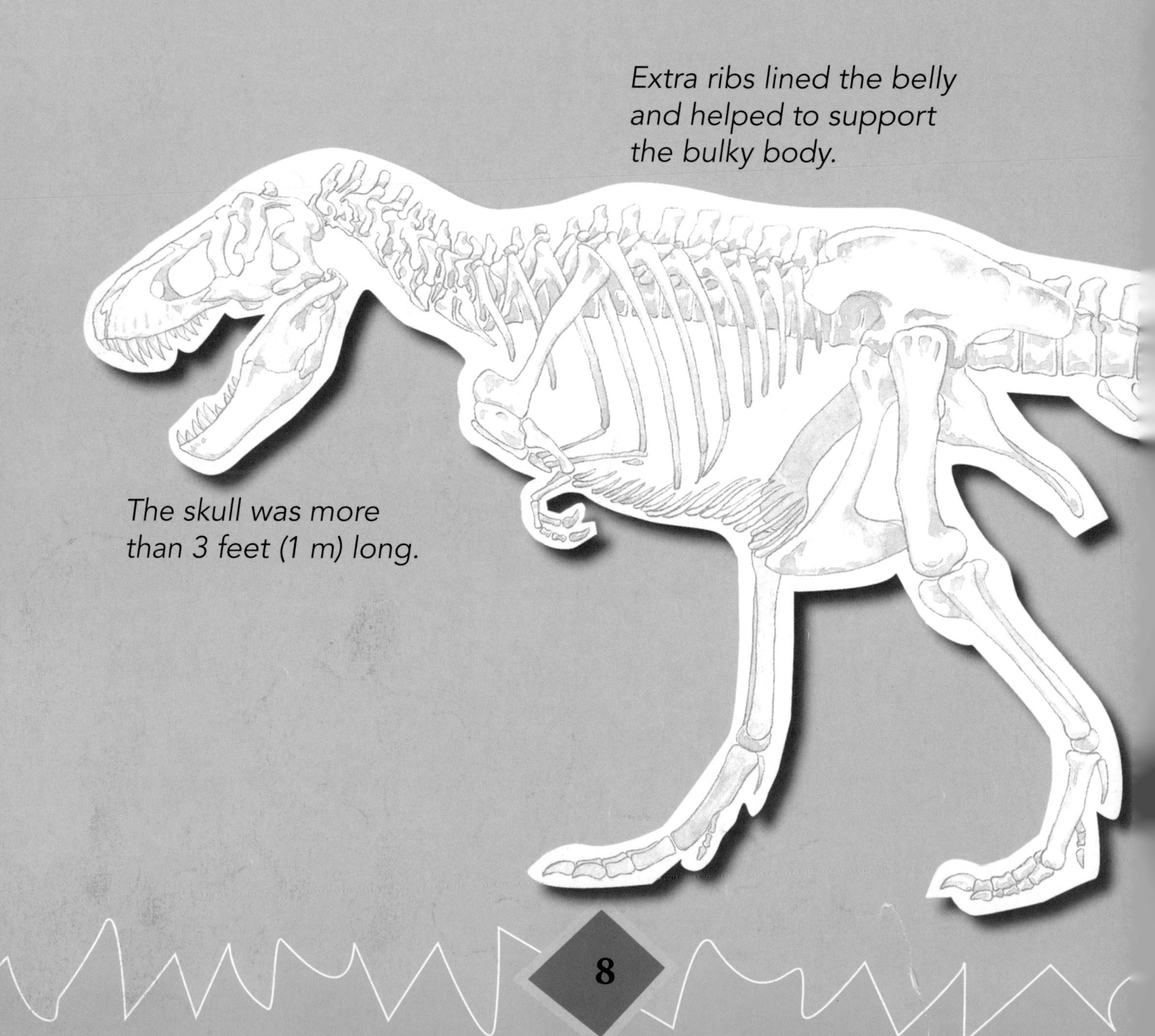

Extra ribs lined the belly and helped to support the bulky body.

The skull was more than 3 feet (1 m) long.

No one knows why the tyrannosaurus and many other big, meat-eating dinosaurs had such tiny arms. Perhaps bigger, heavier arms would have added too much weight to the front of the dinosaur's body.

The tyrannosaurus may have used its arms to help it get up from the ground after sleeping or eating.

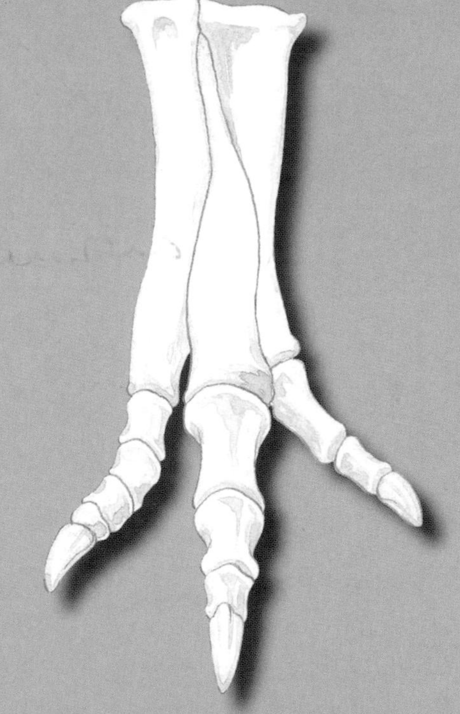

The back feet were big and broad, with three large clawed toes and one small toe at the back.

This dinosaur's huge tail contained as many as 40 sturdy bones.

The tyrannosaurus had long leg and ankle bones, like other fast-running animals.

Some teeth were twice the length of your finger.

A tyrannosaurus in action

An animal the size of the tyrannosaurus probably ate about 165 pounds (75 kg) of meat each day. That's the same as eating more than 40 chickens or several hundred steaks.

There were some very big prey animals that the tyrannosaurus could catch—including plant-eaters such as duckbill dinosaurs and horned dinosaurs. Dinosaur experts argue about how the tyrannosaurus found enough food.

The tyrannosaurus probably couldn't run fast for long, so it would have attacked with surprise.

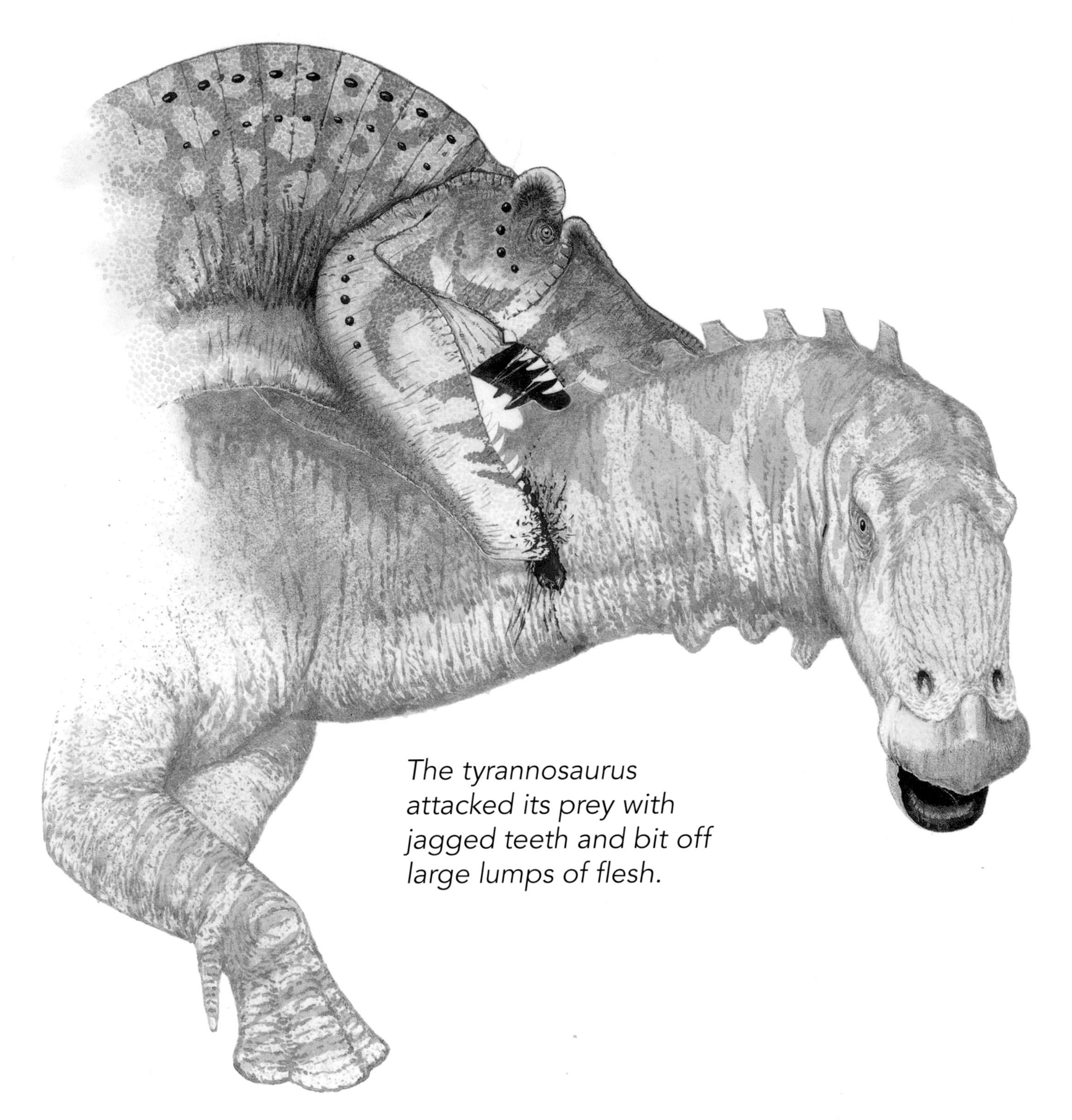

The tyrannosaurus attacked its prey with jagged teeth and bit off large lumps of flesh.

Some think they were skilled hunters who could catch all they needed to eat. Others believe the tyrannosaurus did more scavenging—feeding on animals that were already dead.

We will never know for sure, but the tyrannosaurus probably ate whatever it could find to satisfy its huge appetite.

Young tyrannosaurs

The tyrannosaurs laid eggs from which their young hatched, as other dinosaurs did.

Once a female had laid her eggs, she may have covered them with a pile of plants and dirt to keep them warm. She probably stayed close until they hatched.

Young tyrannosaurs had strong teeth for eating prey, but they were not big enough to catch anything. They probably stayed close to their mom and ate the food that she caught for them.

A mother tyrannosaurus may have guarded her eggs, as crocodiles do today.

These young dinosaurs had growth spurts. Fossils show that a young tyrannosaur gained about 4 pounds (2 kg) a day between the ages of 14 and 18 years.

Teenage tyrannosaurs may have gained an amazing 6,600 pounds (3,000 kg) in just four years!

Albertosaurus

This smaller relative of the tyrannosaurus was still a giant by today's standards. It was as long as two cars.

The albertosaurus was lighter than the tyrannosaurus, so it may have run faster to catch various fast-moving prey.

This is how you say albertosaurus:
al-bert-oh-sore-us

The albertosaurus may have hunted in small packs. Together these dinosaurs could bring down giant plant-eaters such as the sauropods, which were much larger than they were.

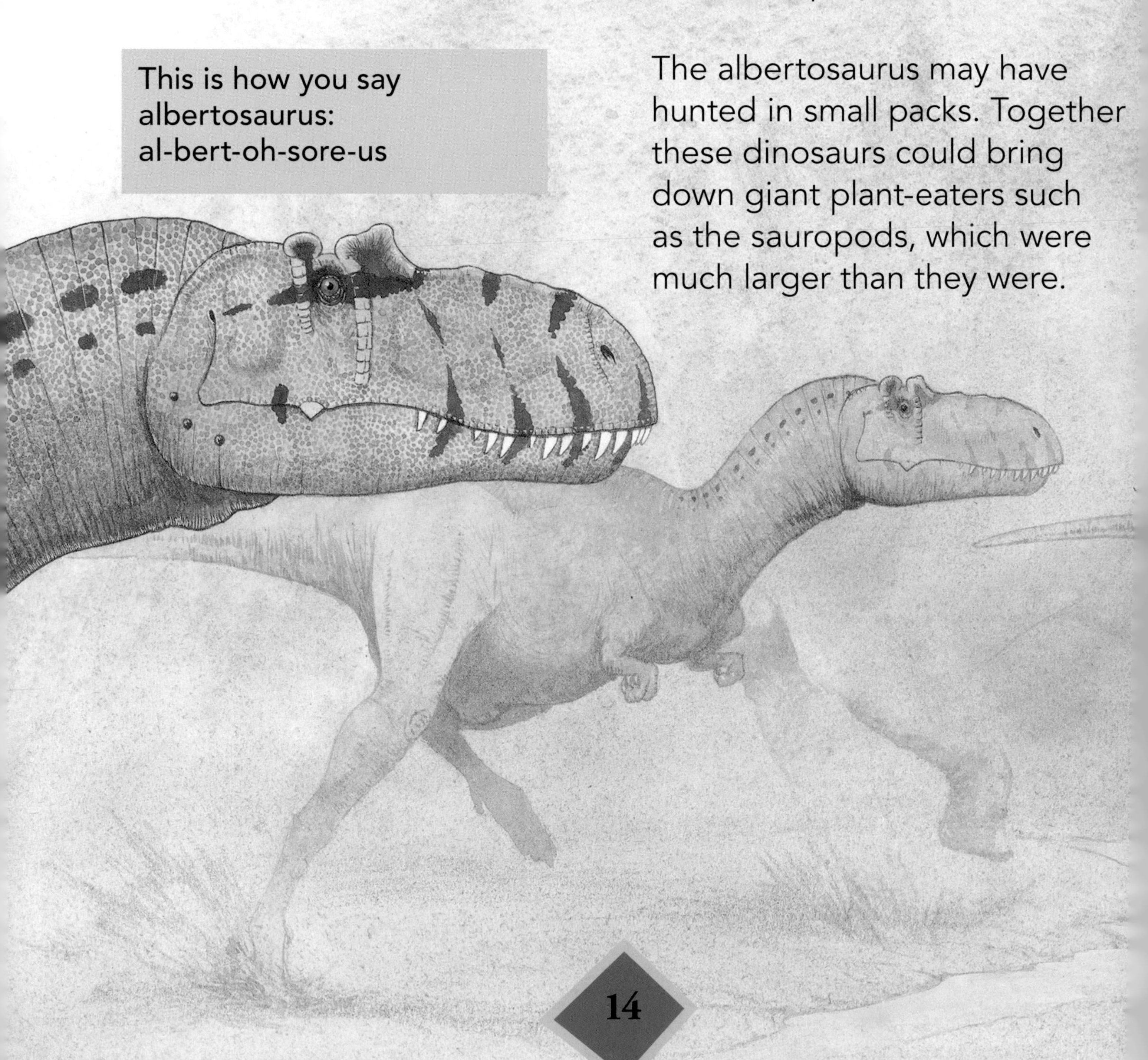

Several hungry albertosaurus dinosaurs could have worked together to hunt big animals.

ALBERTOSAURUS

Group: tyrannosaurids

Length: up to 29.5 feet (9 m)

Lived in: Canada

When: Late Cretaceous, 76–74 million years ago

Giganotosaurus

Fossils of this meat-eating dinosaur suggest that it might have been even bigger than the tyrannosaurus.

The giganotosaurus's head was almost 6.5 feet (2 m) long —that's longer than a human's whole body. The dinosaur may have weighed almost 9 tons (8.2 t); that's more than 100 adult humans!

Every other animal would have run away in terror when the giganotosaurus thundered into view.

GIGANOTOSAURUS

Group: allosaurids

Length: up to 45 feet (13.7 m)

Lived in: Argentina

When: Early Cretaceous, 112–90 million years ago

This is how you say giganotosaurus: jie-gan-oh-toe-sore-us

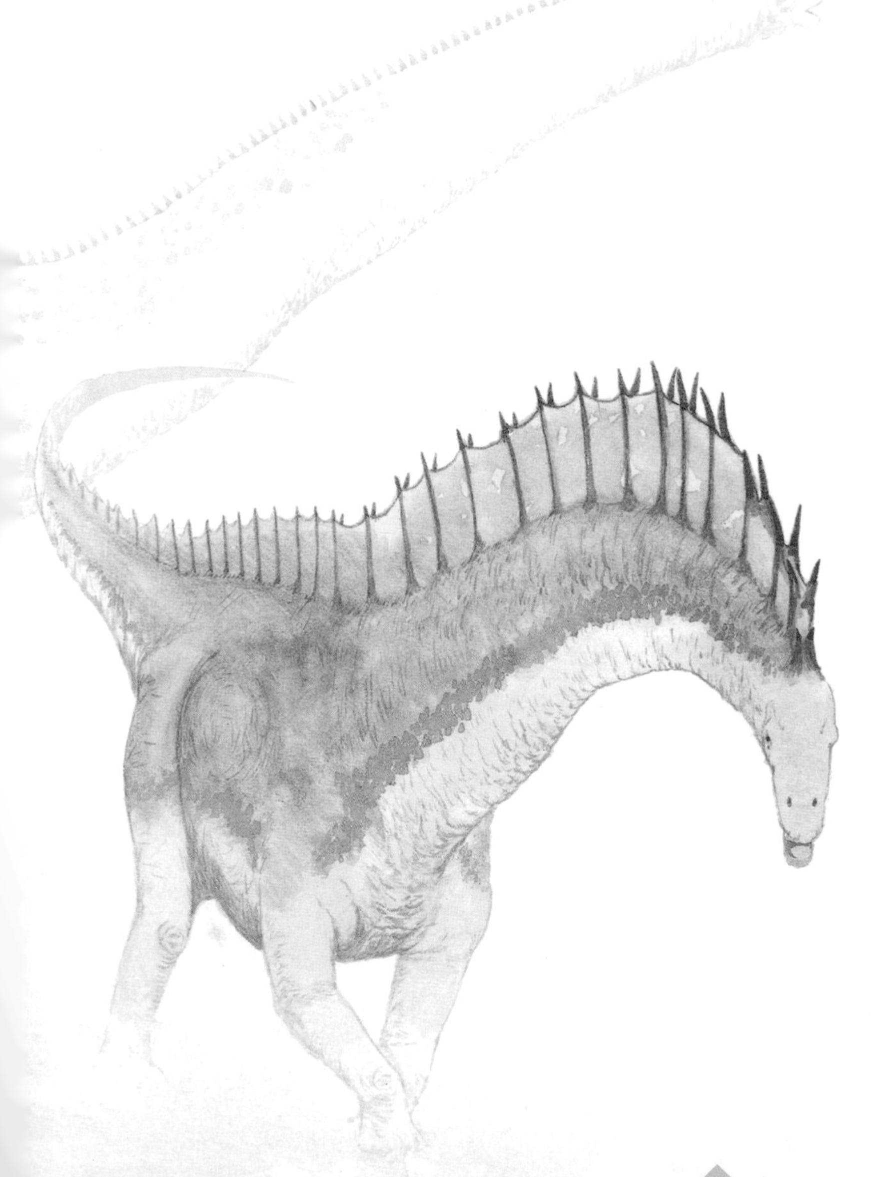

Like the tyrannosaurus, the giganotosaurus had very strong jaws and long, saw-edged teeth. Its arms were very short, but its three-fingered hands had sharp claws to seize prey.

The giganotosaurus belonged to the allosaur group of meat-eating dinosaurs, which lived before the tyrannosaurs.

Yangchuanosaurus

This fierce hunter lived in what is now China. It preyed on large plant-eaters such as the mamenchisaurus, a long-necked sauropod dinosaur.

Like its tyrannosaur cousins, the yangchuanosaurus had dagger-like teeth and sharp claws for attacking prey. Its head was huge, but there were six large holes in its bony skull, which made it lighter.

YANGCHUANOSAURUS

Group: allosaurids

Length: up to 33 feet (10 m)

Lived in: China

When: Late Jurassic, 160–144 million years ago

This is how you say yangchuanosaurus: yang-choo-an-oh-sore-us

The yangchuanosaurus killed with its vicious teeth, but may have also slashed its victim with the claws on its hands and feet.

Allosaurus

There were many allosaurus dinosaurs around in Jurassic times. This hunter even attacked sauropods such as the diplodocus that lived in the same area.

The allosaurus did not always escape unharmed, though. A lash from a sauropod's tail could break its leg or ribs and cripple it.

The fossilized brain of an allosaurus was found in North America.

The brain showed that the dinosaur had a very good sense of smell, but its hearing and sight were not as good.

This is how you say allosaurus:
al-oh-sore-us

The female allosaurus was probably even bigger and more fierce than the male.

ALLOSAURUS

Group: allosaurids

Length: up to 39 feet (12 m)

Lived in: North America

When: Late Jurassic, 153–135 million years ago

Carcharodontosaurus

The name carcharodontosaurus means shark-toothed lizard. The teeth of this dinosaur look as deadly as those of a huge shark.

This meat-eating dinosaur may have been even bigger than the tyrannosaurus, but its bones show that it had a smaller brain.

In 1995, a fossil-hunting expedition found an amazing carcharodontosaurus skull in Morocco. Other fossils of this dinosaur were discovered during the 1920s. These were destroyed by bombs during World War II.

This is how you say carcharodontosaurus: car-car-oh-don-toe-sore-us

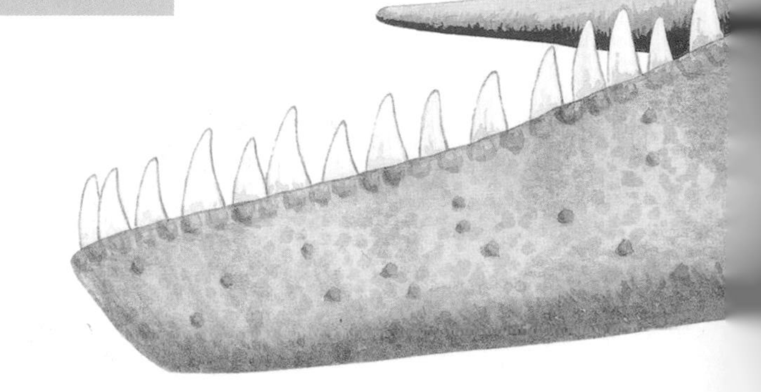

This predator's teeth were 4.7 inches (15 cm) long and its head measured 5.2 feet (1.6 m).

CARCHARODONTOSAURUS

Group: allosaurids

Length: could have been up to 49 feet (15 m)

Lived in: north Africa

When: Late Cretaceous, 98–94 million years ago

Spinosaurus

This animal was very different from any other meat-eating dinosaur. It had tall bones all along its back that were probably covered in skin and looked like a sail. These bones stood as high as an adult human.

The spinosaurus lived in north Africa where it was very hot, and some experts think its sail may have helped it stay cool. Or it may have been a way of attracting a mate or scaring off enemies—the bigger the sail the better.

SPINOSAURUS

Group: spinosaurids

Length: up to 59 feet (18 m)

Lived in: north Africa

When: Late Cretaceous, 95–70 million years ago

This is how you say spinosaurus:
spin-oh-sore-us

The spinosaurus was almost as long as the tyrannosaurus but was lighter and thinner. It had longer arms, too.

Baryonyx

The baryonyx had jaws like a crocodile that were lined with more than 90 sharp teeth.

Its main prey was fish that it grabbed from the water with its long jaws or hooked with its claws.

The baryonyx was big and strong enough to hunt prey on land, too. It could have used the curved claws on its hands to attack its victims.

This is how you say baryonyx:
bar-ee-on-icks

BARYONYX

Group: spinosaurids

Length: up to 32.8 feet (10 m)

Lived in: Europe

When: Early Cretaceous, 125 million years ago

This dinosaur's front legs were only slightly smaller than its back legs, so it could have walked on all fours some of the time.

Bones of this unusual dinosaur were first discovered in 1983 by an amateur fossil hunter in the United Kingdom (UK).

The baryonyx had a big, curved claw on each hand that was nearly 1 foot (30 cm) long—almost as long as your arm!

The end of the dinosaurs?

Most dinosaurs died out about 65 million years ago, along with many other kinds of animals. We can't be sure why, but scientists have several ideas.

One explanation is that a giant lump of rock from space called a meteorite crashed into Earth 65 million years ago. The explosion would have created huge amounts of dust and debris that blocked the sun's rays and changed the world's climate for a long time.

Without sunshine, plants die, so the plant-eating animals starve. Meat-eaters cannot survive without prey, so the predatory dinosaurs died out, too. Scientists have found the remains of a huge crater in Mexico where a giant meteorite may have landed.

Other experts say that dinosaurs had begun to die out long before the meteorite crashed. So the dinosaurs were destroyed by the weather changing on Earth as well as lots of volcanic eruptions. We'll probably never know exactly what happened or why some creatures, such as crocodiles, survived.

Words to know

Allosaurs
A group of large, meat-eating dinosaurs. The allosaurus was an allosaur.

Duckbill dinosaurs
Dinosaurs with a long, flattened beak, like a duck's beak, at the front of the jaws. The lambeosaurus was a duckbill dinosaur.

Fossil
Parts of an animal, such as bones and teeth, that have been preserved in rock for millions of years.

Herbivore
An animal that eats plants. The triceratops was an herbivore.

Herd
A group of animals that usually moves and feeds together.

Horned dinosaurs
Dinosaurs with big, pointed horns on their heads and a sheet of bone called a frill at the back of their heads. The triceratops was a horned dinosaur.

Paleontologist
A scientist who looks for and studies fossils to learn more about the creatures of the past.

Predatory
Describes an animal that hunts and kills other animals.

Prey
Animals caught and killed by hunters such as the tyrannosaurus.

Reptile
An animal with a backbone and a dry, scaly body. Most reptiles lay eggs with leathery shells. Dinosaurs were reptiles. Today's reptiles include lizards, snakes, and crocodiles.

Sauropods
A group of long-necked, plant-eating dinosaurs that includes the largest dinosaurs known.

Scavenging
Feeding on animals that are already dead.

Tyrannosaur
A type of large, meat-eating dinosaur, such as the tyrannosaurus, that attacked plant-eating dinosaurs.

Index